AF556479

# Chandrayaan-3

## INDIA'S GRACEFUL MOONWALK

# Chandrayaan-3

## INDIA'S GRACEFUL MOONWALK

PRIYANKA JAIN

*Published by*

**PRABHAT PRAKASHAN PVT. LTD.**
4/19 Asaf Ali Road,
New Delhi-110 002 (INDIA)
e-mail: prabhatbooks@gmail.com

ISBN 978-93-5562-985-2
**CHANDRAYAAN-3: INDIA'S GRACEFUL MOONWALK**
*by* Priyanka Jain



*Edition*
2026

*Price*
₹ 300 (Rupees Three Hundred Only)

*Printed at*
R-Tech Offset Printers, Delhi

# Author's Note

Why do we explore the moon? What drives us to reach for the stars? These are questions that have puzzled us all for centuries. The moon, with its gentle glow in the night sky, has always been a celestial muse, inspiring poets and scientists alike. However, the reasons for exploring it go beyond mere fascination.

The moon is a treasure trove of mysteries waiting to be unravelled. It holds secrets about the early history of our solar system, Earth's evolution, and the possibility of life beyond our planet. It's a cosmic time capsule; preserving stories and information that can help us understand our own existence better.

The desire to explore the moon is not unique to any one nation or culture. It's a shared human aspiration, a collective yearning to push the boundaries of our knowledge. It's a reminder that, beneath our differences, we are all stardust—connected by the same cosmic forces.

And so, we come to India's remarkable journey to the moon, a journey that has taken the nation to the forefront of lunar exploration. India, a land of diverse cultures, ancient traditions, and a rich history, has emerged as a powerhouse in space technology as well. On the historic day, 23rd August 2023, India's Chandrayaan-3 had a soft touchdown on the lunar surface, marking a turning point in the country's space odyssey.

Chandrayaan-3's achievement is not just about landing on the moon; it's about India's determination to push the boundaries of science and technology. It's about joining the elite club of lunar explorers—the United States, Russia and China. It's also about being the first to make a successful landing near the enigmatic south pole of the moon.

This book will take you through this epic journey and will help you explore the scientific marvels and human dedication that have fueled India's lunar ambitions. It's not a technical manual filled with jargon but a tale that we are proud of. We'll delve into the 'processes and threads' associated with India's Moon agenda, unravelling the purpose, the science, the technology, the history, the efforts, and the challenges.

India's space program, led by the Indian Space Research Organisation (ISRO), is a testament to what humans can achieve when they work together, fueled by curiosity and driven by a vision. Chandrayaan-3 is not an isolated event but a chapter in a grand narrative of discovery and innovation.

So, let us embark on this journey and explore the moon's secrets, celebrate human ingenuity, and dive into the enigma of space. As we navigate the pages of India's odyssey to the moon, let's remember that we are all a part of this cosmic tale. The moon may seem distant but it's a part of our celestial neighbourhood and its mysteries are ours to uncover.

# Contents

# Chapter 1

# Exploring the Moon: A Global Perspective

In the vast expanse of our universe, there exists a celestial body that has captivated human imagination for centuries—the moon. As it hangs in the night sky, sometimes a radiant beacon and at other times a subtle crescent, the moon has been a source of wonder, inspiration, and scientific curiosity. It is a constant presence, yet it remains enigmatic. .

Our journey to explore the moon began in the 1950s, an era when the world was on the cusp of a technological revolution. The desire to reach out to this enigmatic neighbour in the sky sparked a series of missions, each more ambitious than the last. Today, we'll take a step back and understand the types of moon missions that have shaped our quest to unlock the moon's secrets.

## Types of Moon Missions

### 1. Fly-by Missions: Pioneering Lunar Encounters

Imagine a spacecraft hurtling through space, getting ever closer to the moon but never landing . Those were the early explorers, the 'fly-bys.' The first triumph in this category came in January 1959 when the Soviet Union's Luna-1 ventured close to the moon, becoming the first human-made object to do so. Luna-3, later that year, achieved an even more remarkable feat, capturing the moon's first close-up photos. These missions laid the foundation for more complex lunar explorations. Today, fly-bys are employed when the moon serves as a stepping stone to other cosmic destinations.

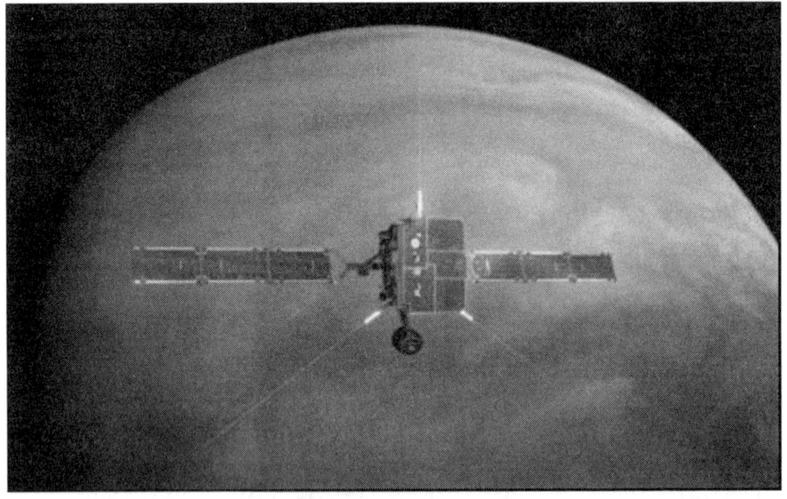

*Solar Orbit Venus Flyby*

### 2. Orbiter Missions: Mapping the Lunar World

Orbiter missions take us closer to the moon, allowing us to study its surface and atmosphere. These spacecraft enter what's known as the 'Lunar Orbit', circling the moon while conducting detailed observations. Over 40 successful orbiter missions have been launched, making this one of the most common forms of lunar exploration. The moon's secrets were gradually unveiled as orbiter missions provided a wealth of data about its features and composition.

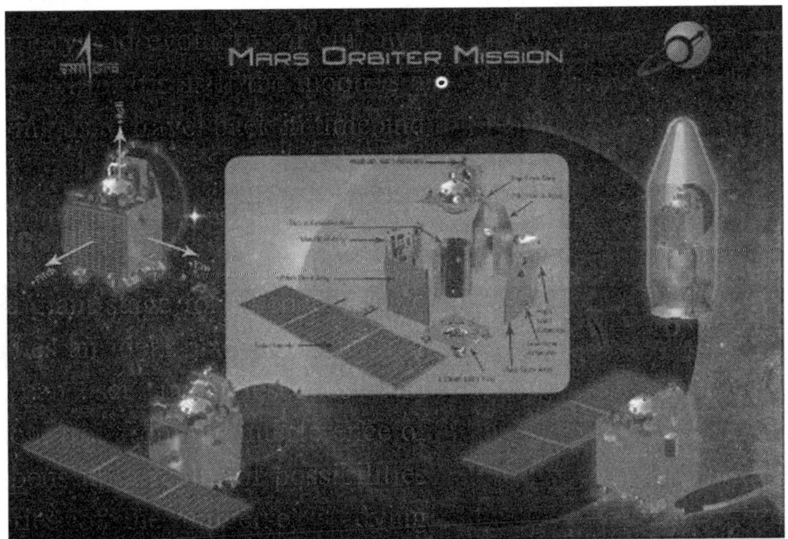


*Mars Orbiter Mission (MOM)*

## 3. Impact Missions: Creating a New Chapter in Exploration

Imagine the primary spacecraft continuing to orbit while a detachable section hurtles towards the moon's surface for a planned crash-landing. This fascinating approach is the hallmark of impact missions. India's Chandrayaan-1 employed this method and its detachable module, the Moon Impact Probe, contributed valuable data during its descent. Impact missions represent a dynamic chapter in lunar exploration, allowing us to gain insights from a closer vantage point.

### 4. Lander Missions: The Triumph of Soft Touchdowns

A spacecraft gently descends onto the moon's surface—a soft landing. This is the pinnacle of lunar exploration. Lander missions are the most complex, and early attempts by the United States and the Soviet Union have faced numerous failures. It wasn't until 1966 that the Soviet Union's Luna-9 achieved the world's first successful lunar landing. The moon's surface, once an enigma, became a platform for scientific investigation.

### 5. Rover Missions: The Moon's Explorers

Small robots equipped with wheels roam the moon's surface, studying its geology and gathering invaluable data. The Soviet Union launched the first successful lunar rover in November 1970. These rovers have expanded our understanding of the moon's composition and dynamics. China's Yutu-2, with its extended operational life, is a testament to the possibilities of lunar exploration.

### 6. Human Missions: The Pioneers

Humans embark on landers and make historic lunar landings. Neil Armstrong's famous words, as he stepped onto the moon in 1969, are etched in history. Following the successful Apollo missions, human exploration of the moon was paused in the early 1970s. The challenges of sending humans to the moon and returning them safely to Earth are immense, but the dream of further lunar exploration continues.

*Human Space Missions*

## Significance of Lunar Exploration

Lunar exploration is like a cosmic treasure hunt, revealing secrets not just about the moon itself but also about our own planet and the vast universe beyond. Let's uncover why these moon missions are so significant, and why they matter not only to scientists but to all of us.

### 1. Unveiling Earth's Early Days

Imagine the moon as a time capsule, preserving stories of how rocky planets like Earth were born. Earth has evolved so much over time that its early chapters are often challenging to decipher. The moon, however, retains this ancient record. It offers glimpses into Earth's infancy when it was a hot, turbulent mess. By studying the moon, we can piece together Earth's own history, like reading a cosmic history book.

## 2. Cosmic Clues from Our Lunar Neighbor

The moon, covered in a layer of dusty soil called regolith, acts like a cosmic detective. It holds clues about space phenomena that have shaped our solar system over billions of years. These clues include meteorites, the solar wind, cosmic rays, and space particles. By examining the moon, we gain insights into the history and evolution of our own planet, including the origins of life. It's as if the moon is a cosmic time machine, allowing us to travel back in time and unlock secrets from our past.

## 3. The Moon as a Scientific Hub

Now, picture the moon as a giant stage for science. As we explore its surface, it becomes an ideal place for scientific investigations. Scientists can set up telescopes and research stations on the moon, free from the interference of Earth's atmosphere. This opens up a realm of possibilities for unravelling the mysteries of the universe, studying extraterrestrial life, and understanding how our bodies function in space. It's like having a front-row seat at the greatest cosmic show.

# Past and Present of Moon Missions across the Globe

## Early Lunar Exploration

The tale of lunar exploration unfolds like an epic. In the mid-20th century, as technology raced ahead, the moon beckoned. The Soviet Union and the United States were the pioneers.

In the late 1950s and early 1960s, both nations embarked on lunar missions. The Soviet Union's Luna program was the first to reach the moon. In 1959, Luna 1 became the first human-made object to venture into lunar space. The climax came in 1966 when Luna 9 made history by successfully becoming the first lunar lander to capture images of the moon's surface and transmit them back to Earth. Humanity had embarked on a new chapter of exploration.

## Post-Apollo Era

After the iconic Apollo missions conducted by NASA in the late 1960s and early 1970s, lunar exploration experienced a brief lull. The monumental Apollo 11 mission in 1969 saw Neil Armstrong and Buzz Aldrin take their historic steps on the moon, achieving significant scientific milestones. While human exploration of the moon paused, robotic missions continued to advance lunar exploration. The Lunar Orbiters and Soviet Luna missions paved the way for modern lunar endeavours.

## Modern Lunar Missions: A Resurgence of Interest

In recent decades, there has been a remarkable resurgence of interest in lunar exploration, fueled by advancements in technology and international collaboration.

- Orbiter missions have become a mainstay, with NASA's Lunar Reconnaissance Orbiter leading the way since 2009. International cooperation, as seen in NASA's Artemis program, aims to bring humans back to the Moon and establish a sustainable presence, heralding a new chapter in lunar exploration and the prospect of expanding human civilization beyond Earth.

- Impact missions have offered the opportunity for more detailed study of the lunar surface, with India's Chandrayaan-1 mission making a significant contribution to this growing body of knowledge. India, through its national space agency, the Indian Space Research Organisation (ISRO), ventured into lunar exploration with the Chandrayaan program.
- China has made remarkable strides in lunar exploration, pioneering the Chang'e program. The Chang'e missions have demonstrated China's prowess in lunar exploration, from orbital missions to successful landings on the lunar surface. The deployment of lunar rovers, such as Yutu-2, has yielded significant scientific discoveries, including revelations about lunar dust depth. The Chang'e missions also signify China's ambitions to explore the Moon's resources for potential use as an energy source on Earth, particularly helium-3.

Lunar exploration isn't just about looking at a shiny object in the night sky. It's about peering into our past, solving cosmic puzzles, and unlocking the doors to a universe brimming with secrets. These moon missions are like pages in a captivating book, and each page turned brings us closer to the thrilling stories of space and time.

# Chapter-2

# Chandrayaan-1: India's First Step towards the Moon

## The Genesis of Chandrayaan-1

India's journey into the cosmos began as an aspiration deeply embedded in its space research endeavours. The inception of space research activities dates back to 1963 when the Thumba Equatorial Rocket Launching Station (TERLS) was established near Thiruvananthapuram. This marked a significant milestone in the study of ionospheric electrojets and related phenomena, initiating India's quest for space exploration.

In 1975, India launched its first satellite, Aryabhata, which carried out scientific experiments in X-ray astronomy, solar neutrons and supra-thermal electron density. This early foray into space research laid the foundation for subsequent advancements. Over the years, scientific research has extended to high-altitude balloons, sounding rockets and satellites and various ground-based facilities have been established for astrophysical, solar and atmospheric research.

India boasts of extensive experience in developing and launching operational spacecraft systems for the survey and management of natural resources, meteorological services and satellite communication. The technologies cultivated in these endeavours became valuable assets for future planetary missions.

The Indian Space Research Organization (ISRO) introduced the Polar Satellite Launch Vehicle (PSLV) and Geosynchronous Satellite Launch Vehicle (GSLV), both capable of embarking on missions to the Moon and nearby planets. India's scientific and technical capacities, coupled with the enthusiasm of its modern scientists, provided the impetus for a lunar mission.

Polar Satellite Launch Vehicle (PSLV): It is a fourth-stage launch vehicle that can place remote sensing satellites into sun synchronous orbit* (SSO). It can also place a small spacecraft into geostationary transfer orbit.

- Size: 44 meters tall
- Mass: 320 tons
- Stages: Four stages with alternating solid and liquid propulsion
- Propellant: The first stage carries 139 tons of solid propellant
- Payload capacity: 600 kg in low earth orbit (LEO) to 1,900 kg in SSO
- History:
    - The PSLV was first successfully launched in October 1994.
    - It was the first Indian launch vehicle to be equipped with liquid stages.

Geosynchronous Satellite Launch Vehicle (GSLV): It is an expendable launch vehicle developed and operated by the Indian Space Research Organisation (ISRO) for the purpose of launching satellites into geosynchronous transfer orbit* (GTO).

Size: 49.13 m tallLift-off

Mass: 420 tons

- Stages: GSLV has three stages with strap-on motors. The first stage uses a 138 tonne S139 solid rocket motor with four liquid engine strap-on motors.
- Payload capacity: GSLV can transport payloads up to 2,500 kg for launches into geosynchronous orbit.
- History:

  GSLV has been used in fifteen launches since 2001.
  - The most recent launch was on March 29, 2018 carrying the GSAT-6A.
  - GSLV is the tallest vehicle operated by ISRO.

## The Birth of a Mission

The idea of an Indian mission to the Moon emerged from discussions within the scientific community. In 1999, the Indian Academy of Sciences initiated the notion, which was further deliberated upon by the Astronautical Society of India in 2000. These discussions laid the groundwork for the formation of a National Lunar Mission Task Force by ISRO.

This Task Force, comprising leading scientists and technologists from across India, evaluated the feasibility of an Indian mission to the Moon. Subsequently, a peer group of over a hundred eminent scientists, representing

*Chandrayaan 1 spacecraft ready for thermovac test*

diverse fields of planetary and space sciences, earth sciences, physics, chemistry, astronomy, astrophysics, engineering and communication sciences, convened in 2003. They unanimously recommended India's commitment to a lunar mission, particularly in light of renewed international interest in lunar exploration in the new millennium.

The Chandrayaan-1 mission became a crucial catalyst for young scientists, providing a unique platform for fundamental research. It was approved by the Government of India in November 2003.

## Scientific Objectives

Chandrayaan-1 set forth an array of scientific objectives, distinctively aimed at high-resolution remote sensing of the Moon. These objectives included:

1. Creating a three-dimensional atlas of the near and far sides of the Moon with high spatial and altitude resolution (5-10 meters).

2. Conducting chemical and mineralogical mapping of the entire lunar surface to determine the distribution of

various elements, including Magnesium, Aluminum, Silicon, Calcium, Iron, Titanium, and high atomic number elements like Radon, Uranium, and Thorium.

Simultaneous photo-geological, mineralogical, and chemical mapping were instrumental in identifying various geological units and inferring the Moon's early evolutionary history. The study of chemical composition aided in determining the stratigraphy and nature of the Moon's crust, offering insights into the formation of our celestial neighbour and the Earth itself.

## Scientific Payloads

Chandrayaan-1 featured an assortment of scientific payloads, each meticulously designed to serve distinct scientific goals:

1. **Terrain Mapping Stereo Camera (TMC):** Designed to map the topography of both near and far sides of the Moon with a spatial resolution of 5 meters and a 20-kilometer swath.

2. **Hyper Spectral Imaging Camera (HySI):** Aimed at obtaining spectroscopic data for mineralogical mapping of the lunar surface with a spectral range of 0.4-0.95 micrometers and a spatial resolution of 80 meters.

3. **Lunar Laser Ranging Instrument (LLRI):** Used to measure the elevation map of the Moon to study basin morphology, stress, strain and lithosphere properties.

4. **High Energy X-ray Spectrometer (HEX):** Covering the hard X-ray region from 30 keV to 270 keV, intended

for studies of natural gamma-rays due to 238U, 232Th, and their decay chain nuclides.

5. **Moon Impact Probe (MIP):** Designed for demonstration of technologies required for impacting a probe at the desired lunar location and for scientific exploration of the Moon's close range.

These payloads were meticulously configured, each with unique technical specifications, designed to provide critical insights into lunar composition, topography and geological history.

## International Collaborations

In addition to the indigenous payloads, ISRO sought collaboration with the international scientific community. Two experiments, Chandrayaan-1 X-ray Spectrometer (C1XS) and Sub keV Atom Reflecting Analyser (SARA), were developed through collaboration between ISRO and the European Space Agency (ESA). This international collaboration brought together the expertise of various countries and expanded the scope of Chandrayaan-1's scientific objectives.

## Launch Vehicle: PSLV-XL

The Polar Satellite Launch Vehicle (PSLV) emerged as the chosen launch vehicle for Chandrayaan-1. The PSLV, developed by ISRO, had proven its reliability and versatility through numerous successful launches, delivering payloads to both Sun Synchronous Orbit

(SSO) and Geosynchronous Transfer Orbit (GTO). PSLV-XL, an upgraded version of PSLV, was selected for Chandrayaan-1, as it could carry the spacecraft into a highly elliptical initial orbit.

## Mission Profile

Chandrayaan-1's journey to the Moon commenced from the Satish Dhawan Space Centre in Sriharikota, India. The spacecraft was launched by PSLV-XL (PSLV-C11) into a highly elliptical initial orbit. A sequence of orbital manoeuvers, propelled by the spacecraft's Liquid Apogee Motor (LAM), gradually raised the apogees and transitioned Chandrayaan-1 toward a lunar trajectory. Once in the vicinity of the Moon, the spacecraft's orbit was meticulously adjusted to facilitate lunar capture, eventually positioning it in a 100-kilometer circular polar orbit around the Moon.

This circular orbit allowed the spacecraft to systematically approach its scientific objectives and initiated the next significant step: the ejection of the Moon Impact Probe (MIP) for a controlled impact on the lunar surface.

Chandrayaan-1's scientific payloads were commissioned, enabling the exploration of the Moon's composition, topography, and geological features over a planned mission duration of two years. Though the operation was supposed to continue for two years, but sadly the radio lost contact with the spacecraft on August 28, 2009 and so the mission had to end.

## Mission Goals and Discoveries

The broader mission objectives encompassed harnessing science payloads, lunar craft and the launch vehicle, in addition to realizing the integration and testing, launching, achieving lunar polar orbit at an altitude of about 100 kilometers, conducting in-orbit experiments, and ensuring communication, telemetry, and data reception for scientific investigation by designated groups of scientists. With the goal of orbiting the moon for two years, the spacecraft had a set of ambitious objectives. It aimed to create a high-resolution lunar atlas and map the moon's surface in incredible detail.

In 2008, Chandrayaan-1 sent a small probe called the MIP to the Moon. Its job was to try out equipment for future missions and to learn about the Moon's very thin air. It eventually landed near the Moon's South Pole and just before it crashed, it found a little bit of water in the Moon's air.

The crown jewel of Chandrayaan-1's achievements was the discovery of water molecules on the moon's surface. This find was groundbreaking because it hinted at the presence of water ice in some of the moon's darkest, coldest corners. The mission even involved a crash landing of the Moon Impact Probe, whose debris helped in the hunt for lunar water. This discovery opened up new possibilities for future lunar missions and even broader space exploration.

## India's Arrival on the Lunar Stage

India's rendezvous with the moon was nothing short of extraordinary. Chandrayaan-1 was a pioneer not just for India but for the entire scientific community. The mission showcased India's potential to undertake complex space missions within a modest budget. The success of Chandrayaan-1 was a resounding global applause for ISRO's professionalism and capability. It displayed India's aspiration to push boundaries, especially in the realm of space science.

India's maiden moon mission had a dual purpose. Though it was about expanding scientific knowledge and upgrading India's technological capabilities, it also offered a stage for young scientists to embark on exciting planetary exploration. India's fascination with the moon extended to its hunt for Helium-3, a potential energy source. Chandrayaan-1 represented India's growing interest in securing its energy future.

The mission's accomplishments resonated globally. India established the Indian Deep Space Network (IDSN) with massive antennas to support lunar and Mars missions. Chandrayaan-1 was not just about reaching the moon; it was about marking India's presence on the lunar stage. Its impact went beyond our planet's satellite; it extended to international prestige and India's commitment to cutting-edge space science.

# Chapter 3

# Chandrayaan-2: A Bold Leap towards the South Pole

India's Chandrayaan-2 mission marked a significant milestone in the country's space exploration journey. Launched on July 22, 2019, this mission was a bold leap towards the moon, with ambitious aspirations and innovative technology at its core.

## The South Pole Quest

One of the most striking features of Chandrayaan-2 was its destination—the moon's South Pole. Why the South Pole? This region had long captured the curiosity of scientists and space enthusiasts. Known for its frigid temperatures and near-constant darkness, the lunar South Pole presented an intriguing challenge. However, it was precisely these challenging conditions that made it a compelling destination.

Scientists believed that the moon's South Pole might hold a valuable resource—water ice. Hidden within permanently shadowed craters, water ice could potentially revolutionize our approach to lunar exploration. Water is not only essential for sustaining life but can also be converted into oxygen and hydrogen, critical for rocket fuel. Chandrayaan-2's primary aspiration was to confirm the presence of water ice in these dark, icy craters.

## Innovations and Technological Marvels

Chandrayaan-2 was more than just an exploration mission; it was a testament to India's technological prowess. The mission comprised three key components: the Orbiter, the Lander (Vikram), and the Rover (Pragyan). These components were a showcase of innovation and cutting-edge technology.

1. **The Orbiter:** Serving as the mission's eye in the sky, the Orbiter was equipped with an impressive array of instruments. It possessed an X-ray spectrometer for analyzing the moon's elemental composition, a solar X-ray monitor for measuring X-ray fluorescence and an imaging spectrometer for mapping minerals and that was just the beginning. The Orbiter featured an infrared spectrometer designed to map water and hydroxyl molecules on the moon's surface, a crucial element of the mission's water-ice search. Furthermore, the Orbiter's high-resolution camera captured breathtaking images of the lunar surface, offering a closer look at our celestial neighbour.

2. **The Lander (Vikram):** The Lander, named Vikram, was engineered for a soft landing on the moon's

surface, a highly challenging feat. It was equipped with instruments to study lunar seismic activity and measure thermal conductivity, providing insights into the moon's internal structure. It also featured a radio spectrometer for investigating the moon's exosphere.

3. **The Rover (Pragyan):** Pragyan, the mission's rover, was a technological marvel. With six wheels and a suite of instruments for soil analysis, it was designed to traverse the lunar surface, collecting valuable data and transmitting it back to the Orbiter. The rover's operation in the moon's harsh conditions, with temperature extremes, was a remarkable technological achievement.

Apart from the primary components, Chandrayaan-2 introduced an indigenously developed landing sensor and navigation system, marking a significant step for India's space program. This system was crucial for precise navigation and landing on the moon.

## The Challenges Faced and Lessons Learned

Chandrayaan-2's journey was not without its share of challenges. Lunar exploration is a complex and demanding endeavour and this mission encountered several hurdles along the way.

1. **The Complex Landing:** Achieving a soft landing on the moon's surface is an intricate and perilous task. The Lander, Vikram, had to navigate a treacherous descent and execute a precise landing. Unfortunately,

the first attempt did not proceed as planned, and contact with Vikram was lost during the descent. It was a moment of heartbreak, but it also underscored the formidable challenges associated with lunar landings.

2. **Extreme Lunar Environment:** The moon's South Pole is an inhospitable environment. The extreme cold and near-constant darkness presented significant challenges to the mission. Chandrayaan-2's instruments and equipment had to withstand these harsh conditions and the engineers had to design them with the lunar extremes in mind.

Despite these challenges, Chandrayaan-2 provided valuable lessons and insights for India's space program and the global space community.

**Resilience:** The setback during Vikram's descent was a stark reminder of the difficulties associated with space exploration. However, it also demonstrated the importance of resilience and determination. ISRO did not give up but continued to communicate with the Orbiter, which continued to send vital data back to Earth.

**Technological Advancements:** Chandrayaan-2 pushed the boundaries of technological innovation. The indigenously developed landing sensor and navigation system were significant achievements, showcasing India's capabilities in space technology. These innovations have broader applications beyond lunar exploration, with implications for future space missions.

**Global Collaboration:** Chandrayaan-2 exemplified global collaboration. The mission carried instruments and payloads from other countries, emphasizing the power of international cooperation in space exploration. This collaboration expanded the mission's scientific scope and highlighted the collective nature of space exploration.

**Inspiration for the Future:** Chandrayaan-2 served as an inspiration to a new generation of scientists and engineers. It demonstrated that even in the face of challenges, ambitious space missions are possible. The mission's setbacks provided valuable lessons for the next generation of space explorers, motivating them to pursue their aspirations in the cosmos.

## The Orbiter's Remarkable Discoveries

While Chandrayaan-2 is often associated with its Lander and Rover, the Orbiter played a pivotal role in expanding our understanding of the moon.

The Orbiter was equipped with a range of scientific instruments that allowed it to conduct detailed observations and experiments from lunar orbit. These instruments provided a wealth of information about the moon's surface and environment.

1. **Terrain Mapping Camera:** The Terrain Mapping Camera was instrumental in creating high-resolution 3D maps of the moon's surface. It provided insights into the moon's topography and helped identify potential landing sites for future missions.

2. **Solar X-ray Monitor:** This instrument measured X-ray fluorescence from the moon's surface, helping to determine the elemental composition of the lunar regolith. This data was vital for understanding the moon's geology and history.

3. **Dual Frequency Synthetic Aperture Radar:** The radar system allowed the Orbiter to peer beneath the moon's surface, revealing the distribution of water and other substances. This was a crucial element of the mission's quest to confirm the presence of water ice.

4. **Imaging IR Spectrometer:** The spectrometer was used to map water and hydroxyl molecules on the moon's surface. It played a pivotal role in detecting the presence of water ice in the Polar Regions.

5. **Dual Frequency Radio Science Experiment:** This experiment involved studying the electron density in the moon's exosphere. It helped understand the moon's extremely thin and tenuous atmosphere.

## Discoveries from Lunar Orbit

1. **Clues about Water Ice:** The Orbiter's findings provided compelling evidence of water ice on the moon. It detected water molecules and hydroxyl ions on the moon's surface, especially in the Polar Regions. This discovery had profound implications for future lunar exploration and potential lunar colonization.

2. **Unveiling the Moon's History:** The high-resolution images captured by the Terrain Mapping Camera

revealed the moon's complex geological history. They showcased the diverse landscapes, including impact craters, volcanic features, and ancient lava plains.

3. **Unravelling the Moon's Mysteries:** The data collected by the Orbiter's instruments allowed scientists to gain insights into the moon's internal structure and geology. This information contributed to a better understanding of the moon's formation and evolution.

Chandrayaan-2's Orbiter, with its remarkable suite of instruments, transformed our knowledge of the moon. It played a pivotal role in confirming the presence of water ice and provided a wealth of data for lunar scientists to unravel the moon's enigmatic past. The Orbiter's discoveries were a testament to the mission's scientific achievements, marking a new era in lunar exploration.

# Chapter 4

# Learning From the Failures

In the realm of space exploration, the path to success is often paved with failures. For every triumph, there are setbacks that serve as valuable lessons. In this chapter, we will delve into the remarkable story of Chandrayaan-3, India's lunar mission that emerged from the ashes of Chandrayaan-2's failure. It's a tale of resilience, determination and the unwavering spirit of scientific exploration.

## A New Dawn after Chandrayaan-2

Chandrayaan-2, India's second lunar mission, had been eagerly anticipated by the entire nation. Launched on July 22, 2019, it was expected to be a resounding success, building upon the achievements of its predecessor, Chandrayaan-1. However, space exploration is fraught with challenges, and on September 6, 2019, the mission encountered a heartbreaking setback when it lost communication with the Vikram lander during its descent to the lunar surface.

Former chief of the Indian Space Research Organisation (ISRO), K Sivan, the man at the helm of these ambitious lunar missions, had this to say during his speech at the ninth convocation ceremony of the National Institute of Technology (NIT) in South Goa: "Just a day after the failure of Chandrayaan-2, we defined Chandrayaan-3. We did not keep it in the corner because of the failure."

This statement underscores the indomitable spirit of ISRO and its commitment to learning from its failures. In this chapter, we will explore the birth of Chandrayaan-3, a mission that arose from the ashes of Chandrayaan-2, and understand how it paved the way for enhanced features and technical advancements.

## The Birth of Chandrayaan-3

In the wake of Chandrayaan-2's setback, there was no room for despondency at ISRO. The very next day, the team began charting the course for Chandrayaan-3. This swift decision showcased ISRO's resolve not to be deterred by failure. Instead, they saw it as an opportunity for growth and improvement.

Approval for this new mission came swiftly from none other than Prime Minister Narendra Modi. Despite the disappointment of Chandrayaan-2's outcome, he recognized the importance of persevering in the pursuit of scientific exploration. It was a testament to his unwavering support for India's space endeavours.

## Enhanced Features and Technical Advancements

As Chandrayaan-3 began to take shape, the ISRO team was determined to make significant advancements and improvements. They were resolute in ensuring that the mistakes and challenges faced during Chandrayaan-2 would not be repeated. The lessons learned from the previous mission served as the foundation for these enhancements.

1. **Improved Communication Systems:** One of the primary issues encountered during Chandrayaan-2 was the loss of communication with the Vikram lander. To

address this, Chandrayaan-3 featured an upgraded and more robust communication system. Redundancies were built in to ensure that any potential glitches in communication could be swiftly mitigated.

2. **Robust Navigation and Landing Systems:** The Chandrayaan-2 mission had encountered difficulties during the landing phase. To enhance the chances of a successful lunar landing, Chandrayaan-3 was equipped with improved navigation and landing systems. These systems were rigorously tested to ensure accuracy and reliability.

3. **Enhanced Payload:** Chandrayaan-3 carried a more extensive payload, including scientific instruments, rovers and other equipment designed to conduct a wide range of experiments on the lunar surface. The additional instruments provided a broader scope for scientific research and data collection.

4. **Redundancy in Critical Systems:** ISRO took a proactive approach in building redundancy into Chandrayaan-3's critical systems. This meant that in case of any technical failures, backup systems could take over, increasing the chances of the mission's success.

5. **Stringent Testing and Simulation:** The team conducted an exhaustive series of simulations and tests to mimic various mission scenarios. This allowed them to identify and rectify potential issues before launch, thus reducing the margin for error.

## Lessons from Chandrayaan-2

Chandrayaan-3 represented not just a fresh start but a quantum leap in terms of technology and preparedness. The setbacks of Chandrayaan-2, particularly the loss of the Vikram lander, served as a critical learning experience. The ISRO team meticulously analyzed the mission's failures and shortcomings, leaving no stone unturned.

These lessons guided the development of Chandrayaan-3 and had a profound impact on the planning and execution of the mission. ISRO's ability to acknowledge its failures and transform them into opportunities for growth showcased the organization's maturity and resilience.

## The Approval and Launch

With enhanced features and technical advancements in place, it was finally time to secure approval for Chandrayaan-3 and prepare for its launch. The Indian government, led by Prime Minister Narendra Modi, displayed unwavering support for the mission, reiterating the country's commitment to space exploration and scientific discovery.

The approval process was swift, reflecting the government's confidence in ISRO's capabilities and the improvements made in the mission's design. As a result, Chandrayaan-3 was slated for launch within a reasonable timeframe, underscoring the urgency and determination of the team.

The day of the launch arrived, and the world watched with bated breath. The successful liftoff marked a significant milestone not only for India but for the global space community. It was a testament to ISRO's ability to rise above adversity and a reminder that setbacks are an inherent part of scientific exploration.

*CH 3 Taking off*

## The Journey to the Moon

Chandrayaan-3's journey to the Moon was marked by meticulous planning and execution. The enhanced communication systems proved to be resilient, maintaining a steady link with the spacecraft throughout its journey. The navigation and landing systems, which had been rigorously tested and improved, guided the mission with precision.

As the spacecraft approached the lunar surface, the tension was palpable. The lessons from Chandrayaan-2 were evident in the cautious and calculated approach. The successful landing of Chandrayaan-3 on the Moon's surface marked a moment of triumph and redemption for ISRO.

Chandrayaan-3's enhanced payload and instruments were put to good use on the lunar surface. The data collected provided valuable insights into the Moon's geology, mineral composition, and other lunar phenomena. It expanded our understanding of our celestial neighbour and contributed significantly to the global body of lunar research.

The mission's success was not just about reaching the Moon but about leveraging the lessons from Chandrayaan-2 to achieve a mission that exceeded expectations. It showcased ISRO's commitment to excellence and its ability to learn from failure, adapt and ultimately triumph.

# Chapter 5

# Major Components and Objectives

## The Crucial Players of Chandrayaan-3: Lander, Rover, and Propulsion Module

Chandrayaan-3's mission was an intricate dance between three key components: the Vikram Lander, the Pragyan Rover, and the Propulsion Module (PM). Each of these components played a pivotal role in ensuring the mission's triumph.

### The Vikram Lander

It had the vital task of making a gentle landing on the Moon's surface. This was no small feat and was often referred to as the 'period of terror' – a nerve-wracking phase lasting around 15 to 18 minutes. What made it even more nail-biting was that during this period, the lander operated autonomously, without real-time commands from ground control. ISRO had left no stone unturned in preparing the lander for its descent. To handle any last-minute glitches, it was equipped with a set of robust and flexible variable-thrust engines.

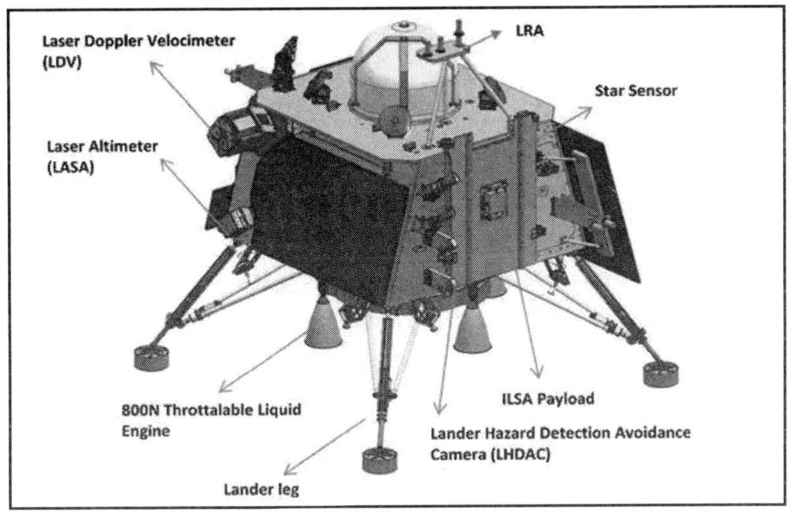


*CH 3 Lander*

Now, what's the challenge? Picture this: The lander had to slow down its horizontal speed from a mind-boggling 1.68 kilometers per second (that's over 6,000 kilometers per hour!) when it was 30 kilometers above the lunar surface. It needed to reduce this velocity almost to zero to achieve a soft landing. In the final phase, the lander had to execute a rather dramatic 90-degree rotation. However, thanks to ISRO's meticulous planning and precise execution, it accomplished this challenging task with finesse.

## The Pragyan Rover

It was another superstar of the mission. Its role was to explore the lunar surface, conduct experiments and gather valuable data. Think of it as a six-wheeled lunar detective. It was armed with a collection of scientific instruments that enabled it to analyze the moon's terrain on the spot. The rover was on a quest to uncover lunar secrets and answer questions about the Moon's composition, the existence of water ice, lunar impacts throughout history and the evolution of the Moon's atmosphere.

## The Propulsion Module (PM)

The Propulsion Module deserves its own round of applause. It played a significant role in transporting the Lander from its initial orbit to the final lunar orbit. You can picture the PM as a box-like structure, with a big solar panel mounted on one side. On top of this module was the Intermodular Adapter Cone, which held the Lander securely. The Propulsion Module also had its scientific payload on board to perform various observations.

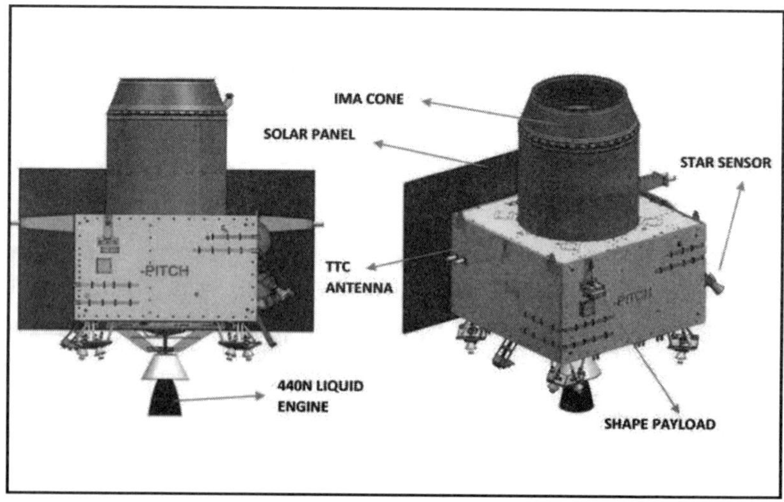


*CH 3 Propulsion Module*

In essence, Chandrayaan-3 was a coordinated team effort, with each component playing its role to perfection. The Vikram Lander's precise descent, the Pragyan Rover's exploration and the Propulsion Module's propulsion- all contributed to the mission's success. It was a cosmic ballet orchestrated by ISRO and the lunar stage was set for Chandrayaan-3 to shine.

## Objectives of the Mission

One of the primary objectives of the Chandrayaan-3 mission was to demonstrate a safe and soft landing on the lunar surface. The need for this demonstration stemmed from the lessons learned during the previous Chandrayaan-2 mission, which had faced challenges during the landing phase.

The success of Chandrayaan-3 in achieving a soft landing is a significant milestone, not just for ISRO but also for the global space exploration community. India joined the ranks of countries with the capability to undertake robotic landings on the Moon, positioning itself as a formidable player in the field of lunar exploration.

The process of achieving a soft landing on the Moon is a complex and highly technical endeavour. It requires precise control and navigation to ensure that the lander reaches the lunar surface with minimal impact. Chandrayaan-3's successful landing was a result of meticulous planning and the integration of advanced technologies.

The lander module (LM) played a crucial role in this phase of the mission. It had been designed and equipped to soft land at a specified lunar site, ensuring a controlled and safe descent. For one lunar day, which is equivalent to 14 Earth days, the lander operated on the Moon's surface, furthering the scientific objectives of the mission.

## The Rover's Journey and Scientific Experiments

While the lander's safe landing was a pivotal achievement, the rover unit, which emerged from the belly of the lander, marked another significant aspect of the Chandrayaan-3 mission. The rover's primary mission was to traverse the lunar surface, conducting scientific experiments and collecting valuable data.

The rover's journey on the Moon was a testament to human ingenuity and technological excellence. Operating for 12 to 14 days on the lunar surface, it ventured into uncharted territory, enabling scientists to gain insights into the Moon's geological and mineral composition. The data collected during the rover's mission promised to enrich our understanding of Earth's celestial neighbour.

The mobility of the rover allowed it to cover a distance of approximately 100 meters on the lunar surface. This distance may seem modest, but it represented a significant accomplishment in lunar exploration. The rover was equipped with a suite of scientific instruments that enabled it to conduct in-situ chemical analysis of the lunar surface. These experiments yielded valuable information, shedding light on the Moon's composition and geological history.

*First movement of Rover*

## An Experiment Called SHAPE

Among the various experiments conducted during the Chandrayaan-3 mission, one of the most intriguing ones was an experiment known as SHAPE (Surface Heat and Physical Experiment). This experiment was designed to study the thermal properties of the lunar surface.

SHAPE aimed to provide insights into the Moon's surface temperature variations, thermal conductivity, and heat flux. Understanding these thermal properties was essential for various reasons. It could inform future lunar missions, help in designing safe and efficient habitats for future lunar explorers, and provide valuable data for scientific research.

The experiment involved the deployment of a series of sensors and instruments on the lunar surface. These sensors measured temperature variations at different depths, helping scientists understand how heat was conducted and distributed on the Moon's surface.

The data collected from the SHAPE experiment promised to offer valuable insights into the Moon's geology, its internal thermal dynamics and even the potential presence of subsurface water ice. These findings had the potential to shape future lunar exploration and the development of sustainable lunar habitats.

## Chandrayaan-3 and the Global Lunar Agenda

The success of the Chandrayaan-3 mission should not be viewed in isolation. Instead, it is part of a broader journey that began with the Chandrayaan-1 mission. Chandrayaan-1 had a profound impact on lunar exploration when it made a surprise discovery, in collaboration with NASA, regarding the presence of water on the Moon.

This groundbreaking discovery ignited the scientific community worldwide, reinvigorating the global lunar research agenda. It brought the Moon back into focus as a crucial destination for scientific exploration and potential resource utilization.

The Chandrayaan-3 mission was a continuation of this scientific journey. It served as a testament to India's commitment to advancing lunar research and exploration. Chandrayaan-3 was not just a technological achievement; it was part of a collective effort by the international space community to unlock the Moon's mysteries.

## Advanced Technologies for a Lunar Mission

To achieve the mission objectives of Chandrayaan-3, a range of advanced technologies were deployed in the lander and other components. These technologies were crucial in ensuring the success of the mission.

1. **Altimeters:** Chandrayaan-3 was equipped with both laser-based and radio frequency (RF) altimeters. These instruments were essential for measuring the lander's altitude above the lunar surface, enabling precise control during descent.

2. **Velocimeters:** The mission featured laser Doppler velocimeters and a Lander Horizontal Velocity Camera to measure the lander's velocity. These instruments played a crucial role in ensuring a safe and controlled landing.

3. **Inertial Measurement:** A combination of laser gyro-based inertial referencing and an accelerometer package provided critical data for navigation, guidance and control.

4. **Propulsion System:** The lander was equipped with an 800N throttle able liquid engine, along with 58N attitude

thrusters and throttle able engine control electronics. These propulsion systems allowed for controlled descent and precise manoeuvering.

5. **Navigation, Guidance and Control (NGC):** The NGC system included powered descent trajectory design and associated software elements. It was responsible for ensuring that the lander followed the planned trajectory during descent.

6. **Hazard Detection and Avoidance:** A Lander Hazard Detection and Avoidance Camera, in conjunction with processing algorithms, helped identify potential hazards on the lunar surface and enabled the lander to navigate around them.

7. **Landing Leg Mechanism:** The landing leg mechanism was crucial for ensuring a stable and safe landing. It had been rigorously tested to withstand different touch-down conditions, ensuring that the lander remained upright on the lunar surface.

## Testing on Earth and in Lunar Conditions

Before Chandrayaan-3's launch, a series of comprehensive tests were conducted to validate the lander's capabilities and the advanced technologies it carried. These tests were instrumental in ensuring that the lander would perform as expected on the lunar surface.

1. **Integrated Cold Test:** This test involved the demonstration of integrated sensors and navigation performance using a helicopter as a test platform. It

simulated various aspects of the mission under controlled conditions on Earth.

2. **Integrated Hot Test:** The integrated hot test aimed to demonstrate closed-loop performance using sensors, actuators, and the NGC system. This test used a tower crane as a test platform, mimicking various aspects of the mission's descent and landing.

3. **Lander Leg Mechanism Performance Test:** To ensure that the lander's leg mechanism could withstand different touch-down conditions, tests were conducted on a lunar simulant test bed. These tests simulated the lunar surface's conditions and verified the lander's stability.

*Integrated Module*

# Chapter 6

# The Journey to the Lunar Dawn

The journey of Chandrayaan-3 to the Moon was a remarkable odyssey, filled with scientific challenges and technological triumphs. This chapter unfolds the stages that led the mission to its successful landing on the lunar surface. Before digging deep into the journey, let's have a look at a brief chronological timeline of this mission.

## 1. Preparations on Swing

**July 6, 2023**: ISRO excitedly shares Chandrayaan-3's launch date of July 14, from Sriharikota's secondary pad.

**July 7, 2023**: The vehicle electrical assessments are successfully completed.

**July 11, 2023**: ISRO gears up with a 24-hour launch rehearsal, replicating the launch procedure.

## 2. Lift-off and Celestial Orbits

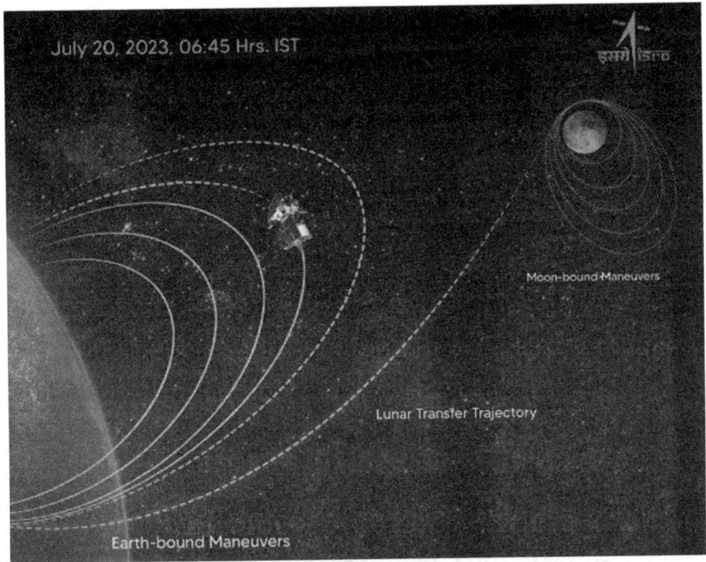


*CH3 July 20, 2023*

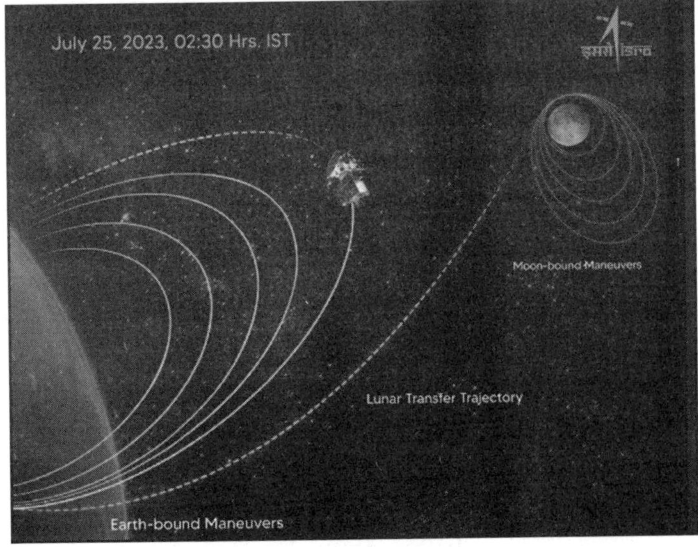


*CH3 July 25, 2023*

**July 14, 2023**: A historic moment as Chandrayaan-3 gracefully soars into the cosmos with the LVM3 M4 vehicle, reaching its designated orbit.

**July 15, 2023**: The mission takes its first steps in space with the first orbit-raising manoeuver, reaching 41,762 km x 173 km.

**July 17, 2023**: In a choreography of orbits, the second manoeuver positions Chandrayaan-3 at 41,603 km x 226 km.

**July 22, 2023**: The dance continues with the third manoeuver, guiding the spacecraft to 71,351 km x 233 km.

**July 25, 2023**: Fine-tuning the path with an additional orbit-raising manoeuvre .

## 3. Gateway to the Moon

**August 1, 2023**: A crucial milestone – Chandrayaan-3 is gently placed in translunar orbit, precisely at 288 km x 369,328 km.

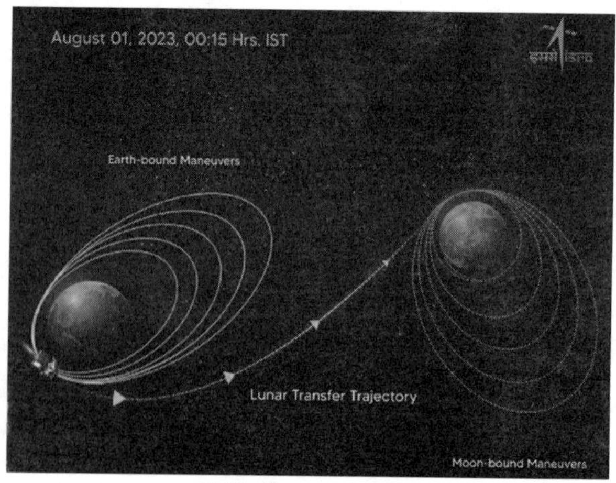


*CH3 August 1, 2023*

**August 5, 2023**: A masterful lunar orbit is achieved, settling at 164 km x 18,074 km.

## 4. Precise Orbit Tuning

**August 6, 2023**: Minor adjustments tweak the lunar orbit to 170 km x 4,313 km.

**August 9, 2023**: The mission maintains its course with Chandrayaan-3's trajectory set at 174 km x 1,437 km.

**August 14, 2023**: A well-timed adjustment positions the orbit at 150 km x 177 km.

**August 20, 2023**: The spacecraft's orbit is finely tuned, reaching the farthest and nearest lunar points at 134 km x 25 km.

## 5. All Set to Land on the Moon

**August 17, 2023**: The heart-pounding moment arrives as the landing module, including the Vikram Lander and Pragyan Rover, separates from the propulsion system.

**August 18, 2023**: 'De-boosting' commences, gently lowering the landing module's orbit to 113 km x 157 km.

**August 20, 2023**: Chandrayaan-3's orbit is adjusted to a precise 134 km x 25 km.

## 6. The Lunar Touchdown

**August 23, 2023**: The curtain rises for the anticipated lunar touchdown, initiated at 5:47 pm IST. A soft landing is planned for 6:04 pm IST.

## 7. Exploration and Discoveries

**August 24, 2023**: The Pragyan Rover embarks on its lunar exploration mission, prepared to unveil the Moon's well-guarded secrets.

**August 30, 2023**: Pragyan confirms the presence of sulphur on the lunar surface, employing advanced Laser-Induced Breakdown Spectroscopy (LIBS).

**September 3, 2023**: The rover enters sleep mode after successfully completing its assignments, gearing up for the impending lunar night.

## 8. Artistic Visions and Challenges

**September 15, 2023**: The Lander Position Detection Camera (LPDC) transforms the Moon into art, capturing a stunning image.

**September 22, 2023**: Tensions rise as the mission faces an uncertain moment, with the lander and rover missing their wake-up calls. By September 28, hopes for further surface operations begin to wane.

This journey of Chandrayaan-3 showcases the marvels of scientific precision and the enigmatic world of the Moon, all thanks to ISRO's unwavering dedication and ingenuity.

# Launch Preparations and the LVM3 M4 Launcher

The Chandrayaan-3 mission was launched using the formidable LVM-3, a launch vehicle that had already proven its reliability by undertaking two commercial launches for the UK-based company OneWeb. During these missions, 36 satellites were launched into low Earth orbit, establishing the vehicle's credibility. However, Chandrayaan-3 was in a different league altogether.

LVM-3 is configured as a three-stage vehicle, consisting of two solid strap-on motors (S200), one liquid core stage (L110) and a high-thrust cryogenic upper stage (C25). What made this launch even more significant was that ISRO seized the opportunity to use LVM-3 to prepare for future human missions, specifically the Gaganyaan project. For Chandrayaan-3, ISRO employed human-rated solid strap-on motors. The success of this launch gave ISRO the confidence to pursue both unmanned and manned missions for the Gaganyaan project using the same rocket.

## The Elliptic Parking Orbit (EPO) and Lunar Arrival

The Chandrayaan-3 mission's journey to the Moon was a meticulously planned and executed process, involving multiple stages and precise manoeuvers. On July 14, 2023, ISRO's LVM-3 placed the Chandrayaan-3 module in an

Elliptic Parking Orbit (EPO) at an altitude of 180 km. The EPO had dimensions of 36,500 km x 180 km, marking the starting point of the lunar journey.

The journey from Earth to the Moon was no small feat; covering a distance of approximately 384,400 km. ISRO undertook a series of 'Earth orbit-raising' manoeuvres to propel the craft further into space. These manoeuvres were essential for ensuring that the spacecraft would reach an altitude of about 100,000 km above the Earth's surface.

One of the distinctive features of ISRO's lunar missions was the innovative approach to traversing the vast distance to the Moon. While major space powers like the US and Russia could reach the Moon in just four to five days, ISRO took a different path due to its limited rocket launcher capabilities. ISRO needed approximately 40 days to reach the Moon, but this challenge was met with ingenuity.

The process of reaching the Moon involved three main stages: Earth orbit manoeuvers, translunar injection, and lunar orbit manoeuvers. These complex stages were carried out successfully, and the lander separation from the PM occurred as planned. The lander then entered an orbit closer to the Moon, bringing it within reach of the lunar surface.

## Chandrayaan-3 Mission Profile

The mission profile of Chandrayaan-3 is a testament to ISRO's precision and dedication. The spacecraft underwent a series of orbit-raising manoeuv res, gradually approaching the Moon.

On July 14, the Chandrayaan-3 module was placed in an EPO at an altitude of 180 km. Subsequent orbit-raising manoeuvres took place on July 15, 16, 18, 20, and 25, gradually increasing the altitude of the craft. These manoeuvres positioned the spacecraft at an altitude of 127,609 km x 236 km.

On August 1, 2023, the critical translunar injection was executed. This manoeuvre propelled Chandrayaan-3 away from Earth's orbit, initiating its journey towards the Moon. The manoeuvre required precision to escape Earth's gravitational force.

August 5 marked a pivotal moment in the Chandrayaan-3 mission with the successful Lunar Orbit Insertion (LOI). This manoeuvre resulted in an orbit of 164 km x 18,074 km. Following the LOI, the mission underwent four more planned orbital manoeuvres to position the spacecraft at an approximate distance of 100 km from the Moon's surface.

On August 6, the spacecraft entered a 170 km x 4,313 km lunar orbit. The orbit was further reduced to 174 km x 1,437 km on August 9. On August 14, the spacecraft was in a 151 km x 179 km orbit, and finally, on August 16, it reached an orbit of 153 km x 163 km.

On August 17, the LM successfully separated from the PM, and de-boosting manoeuvres occurred on August 18. These manoeuvres placed the LM in a lunar orbit of 113 km x 157 km. Subsequent de-boosting on August 20 brought the spacecraft to an orbit of 25 km x 134 km.

The climax of the mission unfolded on August 23, 2023, around 5:45 p.m., with the powered descent phase. It was the most critical part of the entire mission, involving a descent to the lunar surface over approximately 15 to 18 minutes. This phase was autonomous, with ground control having no direct involvement. The system's software had been carefully prepared and the spacecraft was trained to handle any last-minute issues.

The powered descent had four broad phases: Rough Braking Phase, Attitude Hold Phase, Fine Braking Phase and Terminal Descent Phase. One of the most challenging aspects was reducing the lander's horizontal velocity from over 6,000 km/h at a height of 30 km from the lunar surface to nearly zero for a soft landing.

The chosen landing site was near the Moon's South Pole at 70 degrees latitude, specifically at 4 km x 2.4 km, with coordinates of 69.367621°S and 32.348126°E. This decision was not made lightly, as the South Pole area presented unique challenges. The region had large craters and an uneven surface, making it a formidable landing spot. Temperatures in this region could plummet to as low as -230°C, requiring the craft and its sensors to withstand such extreme cold.

The selection of this challenging landing site was deliberate, as ISRO aimed to explore the water ice deposits on the Moon. While the South Pole presented challenges, it also held the promise of important discoveries. ISRO's decision to operate in this region was a testament to its commitment to pushing the boundaries of lunar exploration.

## Communication Challenges and Collaboration

Communication was a critical aspect of the mission, and ISRO had to overcome challenges related to keeping the spacecraft connected with ground stations. While ground stations located on Indian soil played a vital role in tracking various satellite missions, deep space missions required a different approach.

The size of the antenna was a significant factor in maintaining communication with spacecraft on deep space missions. Traditional antennas were designed for electromagnetic waves, which have a relatively long wavelength. ISRO had to collaborate with international partners, including NASA and ESA, to ensure continuous tracking and communication.

ESA's ground stations in Kourou, Goonhilly and New Norcia, along with NASA's DSN, provided telemetry and tracking coverage during the powered descent phase. These collaborative efforts were crucial for the success of the mission, demonstrating the international cooperation that underpins space exploration.

The journey to the lunar dawn was marked by meticulous planning, precise execution and collaboration with international partners. Chandrayaan-3's successful arrival on the lunar surface marked a significant achievement for ISRO and a milestone in India's space exploration journey. The challenges of the journey were met with innovation and determination, setting the stage for the scientific discoveries and achievements to come.

# Chapter 7

# The Lunar South Pole: A Challenging Destination

The choice to explore the lunar South Pole is not merely a matter of convenience; it's driven by a profound understanding of the moon's mysteries and a pursuit of scientific discoveries that promise to reshape our understanding of the lunar surface.

## Why Exploring the Lunar South Pole Matters

Historically, lunar exploration has primarily focused on the moon's equatorial region, positioning missions just a few degrees north or south of the lunar equator. However, with Chandrayaan-3, India ventured into uncharted territory by targeting the lunar South Pole, an area that presents unique challenges and extraordinary rewards.

NASA's Surveyor 7 set a historical precedent when it achieved a lunar touchdown near the 40-degree south latitude on January 10, 1968. This mission, though remarkable, barely scratched the surface of what the lunar South Pole had to offer.

ISRO learned from past experiences and addressed the intricacies posed by the lunar South Pole. The mission involved significant improvements in software and hardware, including modifications to the lander's thrusters, enhanced soft-landing sequences and other technical upgrades. Notably, ISRO reconfigured the lander with four thruster engines instead of five, strengthened landing legs, expanded solar panels and increased fuel capacity. These enhancements were critical to ensure Chandrayaan-3's resounding success.

Global interest in the lunar South Pole aligns with the international trend of exploring this enigmatic region. NASA's Artemis III mission, scheduled for 2025, is a significant endeavor with a primary focus on the lunar South Pole. Chandrayaan-3's ambitious quest is part of this larger scientific narrative.

The South Pole-Aitken basin, an immense crater, cradles the lunar South Pole and holds a treasure trove of geological insights. Researchers believe that this region may contain materials from the moon's deep crust and upper mantle, shedding light on its geological history and evolution.

The presence of water ice at the lunar South Pole adds to its allure. Water is a valuable resource in space exploration, offering the potential for sustainable lunar bases and the ability to generate rocket fuel on the moon itself. The South

Pole boasts of more area in permanent shadow and colder temperatures, making it a promising location for water ice accumulation.

## Tough Conditions, but Big Rewards

Exploring the lunar South Pole comes with a unique set of challenges and equally significant rewards. The decision to land in this region was driven by the tantalizing prospects of scientific discovery and resource utilization.

The lunar South Pole is notorious for its rugged, cratered surface, freezing temperatures, lack of light, and communication difficulties. Attempts to land in this region have been met with substantial complexities and hurdles, making it one of the most challenging areas for lunar missions.

However, the potential rewards are immense. Scientists have long been aware of the presence of frozen water in the lunar South Pole. Identifying these frozen water reservoirs is a significant gateway for further space exploration within our solar system. Water is a critical resource in space, with applications ranging from supporting human missions to serving as a potential source of rocket fuel.

Navigating the complex terrain, darkness, extreme temperatures and communication challenges at the lunar South Pole demands precision and innovation. Chandrayaan-3's mission is dedicated to overcoming these obstacles to unlock the mysteries of the moon's southern polar region.

## Picking the Right Spot for Chandrayaan-3

Selecting the perfect landing site for Chandrayaan-3 required a careful balance between scientific goals, terrain challenges and resource potential. ISRO's decision to land at a specific location near the lunar South Pole was driven by several factors.

The lunar South Pole offers a wealth of scientific insights due to its potential to contain materials from the moon's deep crust and upper mantle. This makes it an exciting location for geological exploration. Water ice has been detected at both lunar poles, but the South Pole is believed to have more due to its larger area in permanent shadow and colder temperatures.

ISRO chose a landing site near the lunar South Pole that was strategic for several reasons. This location was positioned at 70 degrees south latitude, offering a balanced compromise between challenging terrain and the promise of valuable discoveries. It is an area where sunlight is limited, providing a unique opportunity to explore both shadowed and illuminated regions, including craters.

The Chandrayaan-3 mission represents a significant step toward understanding lunar geology, surface processes, and the potential for resource utilization. As the spacecraft navigates these treacherous terrains and faces extreme conditions, the rewards in terms of scientific insights and space exploration advancements are immeasurable. Chandrayaan-3's journey to the lunar South Pole is an audacious endeavor that has the potential to reshape our understanding of the moon as well as our capabilities in space exploration.

## Scientific Importance

- The examination of temperature fluctuations and the identification of ice on the Moon are pivotal in enhancing our comprehension of lunar geology and surface dynamics.
- The exceptional lighting conditions found in the Polar Regions provide a valuable opportunity to analyze ancient impact craters and preserve invaluable historical data.
- These investigations play a crucial role in unravelling the Moon's geological past, the evolution of the Solar System and the potential for harnessing lunar resources.

# Chapter 8

# Chandrayaan-3's Arrival and Lunar Exploration

Chandrayaan-3's mission to the lunar South Pole marked a pivotal moment in India's space exploration endeavors. With the rover and lander in tow, the mission was poised to deliver groundbreaking insights and invaluable data over its relatively short, yet action-packed, mission duration.

## Landing and Rover Deployment

The journey began with a spectacular lunar touchdown, marking a significant milestone in Chandrayaan-3's mission. On 3 September, the rover, nestled within the lander, was carefully lowered to the lunar surface. Once safely on the moon, the rover set about its tasks with precision and purpose. It executed a brief 'hop,' ascending 40 cm off the lunar surface and translating a similar distance laterally. This seemingly small 'hop' was a giant leap in demonstrating capabilities that could be pivotal in potential future lunar sample return missions. The instruments and the rover deployment ramp were retracted during the hop and redeployed afterwards, underscoring the ingenuity and adaptability of Chandrayaan-3's technology.

The rover's actions were carefully orchestrated as it embarked on its scientific tasks, setting out to uncover lunar secrets and provide answers to longstanding questions about the moon's geology, surface composition and potential resources.

## Scientific Payloads and Data Collection

The rover's mission was characterized by meticulous data collection, analysis and transmission. Over the course of its operational period, the rover's scientific payloads played a pivotal role in gathering invaluable data. Among these payloads was the Radio Anatomy of Moon Bound Hypersensitive ionosphere and Atmosphere (RAMBHA). Equipped with a Langmuir probe (LP), RAMBHA measured the near-surface plasma density, including ions and electrons, and monitored its changes over time.

Chandra's Surface Thermo-Physical Experiment (ChaSTE) was another crucial payload on the lander module, designed to measure the thermal properties of the lunar surface in the near-polar region. Instrument for Lunar Seismic Activity (ILSA) aimed to measure seismic activity around the landing site while delineating the structure of the lunar crust and mantle. The LASER Retroreflector Array (LRA) was a passive experiment, contributing to a better understanding of the moon's dynamics.

On the rover module, two vital payloads were Laser-Induced Breakdown Spectroscope (LIBS) and Alpha Particle X-ray Spectrometer (APXS). LIBS conducted qualitative and quantitative elemental analysis, helping to derive the chemical composition and infer mineralogical details of the lunar surface. APXS determined the elemental composition of lunar soil and rocks around the landing site, offering insights into its mineral composition.

**Lunar far side area as imaged from the Lander Hazard Detection and Avoidance Camera (LHDAC) onboard Chandrayaan-3 on August 19, 2023**

## The Mission's Estimated Duration

Chandrayaan-3's mission was remarkable not only for its scientific objectives but also for the mission duration and the conditions it faced. The lander and rover were expected to operate for just one lunar daylight period, equivalent to 14 Earth days. This relatively short operational window was due to the challenges posed by the lunar night, where temperatures could plummet to a bone-chilling -120°C (-184°F).

On 3 September, as the lunar night approached, the rover was carefully placed into sleep mode to conserve energy and protect it from the extreme cold. Its batteries were charged, and the receiver was left on, preparing for the impending lunar night. The rover's data collection activities were completed, and the accumulated data was successfully transmitted to Earth via the lander.

*Pre and post hop ramp images by Lander*

However, as the lunar night continued on 22 September, both the lander and rover missed their wake-up calls. By 28 September, neither had responded, diminishing hopes for further surface operations. Chandrayaan-3's mission was an ambitious and daring endeavor, one that had provided a wealth of data and valuable insights during its brief lunar sojourn.

The mission life of the propulsion module extended to a potential six months as it operated the experimental payload in lunar orbit, paving the way for future interplanetary missions. Chandrayaan-3's journey to the lunar South Pole was a testament to India's innovation and determination, offering a glimpse into the possibilities of future lunar exploration and space missions.

# Chapter 9

# The Team behind the Triumph

The remarkable success of Chandrayaan-3 could not have been possible without the brilliant minds, dedicated agencies and collaborative efforts that brought this ambitious lunar mission to fruition.

## The Brilliant Minds at ISRO

1. **S. Somanath (ISRO Chairman):** At the helm of India's ambitious Moon mission and other space ventures is ISRO chief, S. Somanath. He's the one leading the way, not only for Chandrayaan-3 but also for missions like Gaganyaan and the Sun mission Aditya-L1. Before taking the top spot at ISRO, he was in charge of important centers that develop rocket technology, like the Vikram Sarabhai Space Centre and the Liquid Propulsion Systems Centre. Somanath, who became ISRO's chairman in 2022, played a vital role in making Chandrayaan-3 happen. His mechanical engineering background and expertise in rocket design

*ISRO Chairman S. Somnath*

is key to understanding the mission's history and its exciting journey into space.

2. **P. Veeramuthuvel (Project Director, Chandrayaan-3):** P. Veeramuthuvel has been leading India's latest

*P. Veeramuthuvel (Project Director, Chandrayaan-3)*

moon-landing mission since 2019. Before this lunar mission, he had worked as a deputy director in the Space Infrastructure Programme Office at ISRO headquarters. He's highly skilled in technical matters and previously played a significant role in Chandrayaan-2. In that mission, he was the go-to person for discussions with NASA. Veeramuthuvel hails from Tamil Nadu's Villupuram and is an alumnus of the Indian Institute of Technology in Madras (IIT-M).

3. **S. Unnikrishnan Nair (Director, Vikram Sarabhai Space Centre):** S Unnikrishnan Nair is the head of Vikram Sarabhai Space Centre (VSSC) at Thumba in Kerala's Thiruvananthapuram district. He and his team

*Dr. S Unnikrishnan*

are responsible for the key functions of the crucial mission. The Geosynchronous Satellite Launch Vehicle (GSLV) Mark –III, which was renamed as Launch Vehicle Mark-III rocket, was also developed by the Vikram Sarabhai Space Centre (VSSC).

4. **B. N. Ramakrishna (Director, ISTRAC):** Head of deep space communications, B. N. Ramakrishna brought his extensive knowledge of navigation and orbit determination of spacecraft to the mission. His expertise in ensuring effective communication with lunar probes was vital.

5. **S. Mohana Kumar (Mission Director, Chandrayaan-3):** As a mission director closely associated with various LVM3 missions, S. Mohana Kumar's experience was crucial in managing the Chandrayaan-3 mission.

*Chandrayaan 3 Team ISRO*

6. **M. Sankaran (Director, U.R. Rao Satellite Centre):** M. Sankaran played a pivotal role in designing, developing, and realizing all ISRO satellites, including Chandrayaan-3. His responsibilities also extended to testing the strength of the Vikram Lander.

7. **Kalpana K (Deputy Project Director, Chandrayaan-3):** Kalpana K, an engineer with a diverse background, brought her experience from previous missions, including Chandrayaan-2 and Mars missions, to her role as Deputy Project Director for Chandrayaan-3.

## Women of the Mission

Women have made invaluable contributions to ISRO programs, and the Chandrayaan-3 mission is no exception. A remarkable team of over 100 women scientists and engineers played pivotal roles in various aspects of CH-3,

*Women of CH 3*

from its conceptualization to execution. These exceptional women led the way in spacecraft configuration, managed teams, handled assembly, integration and testing of the spacecraft and set up the ground segment for CH-3 mission operations.

ISRO boasts a diverse workforce, with women making up between 20 and 25 percent of its 16,000-plus employees. Over 100 talented women scientists and engineers actively participated in the Chandrayaan-3 mission, which culminated in India's historic lunar rover landing near the moon's South Pole on August 23.

During the launch and landing on August 23, many women scientists were present in the control room, exemplifying their significant roles in this achievement. Prime Minister Narendra Modi engaged in a conversation with these remarkable women, lauding their essential contributions to India's successful lunar landing mission.

Among the prominent leaders of the mission was Deputy Project Director Kalpana Kalahasti. Her extensive experience includes roles in India's second lunar mission and the Mars orbiter mission. Kalahasti is a satellite specialist responsible for overseeing advanced imaging devices that have enabled ISRO to capture high-resolution images of the Earth's surface.

Reema Ghosh, another trailblazing woman in the mission, is a robotics specialist who played a key role in the development of the "Pragyan" rover currently exploring the lunar surface. During the Prime Minister's visit, she expressed her excitement, saying, "For me, Pragyan is like a baby taking its first steps on the moon. It's a wonderful experience to see the rover in action on the lunar surface for the first time,". These dedicated women scientists and engineers have been crucial in advancing India's space exploration endeavors and setting new milestones in lunar exploration.

## Contribution of the Major ISRO Centres

The major ISRO centers that played pivotal roles in shaping and executing the CH-3 mission are as follows:

### 1. U R Rao Satellite Centre (URSC), Bengaluru

- URSC serves as the lead center for spacecraft conceptualization, design, development, testing, and operations.
- It meticulously designs, tests, and certifies spacecraft structures.
- URSC provided the thermal protection system, crucial for extreme temperature conditions in space.
- They were responsible for developing various mechanisms like payload systems, Rover ramp deployment, antenna deployment, and Lander's legs.
- The electronic subsystems of the satellite are powered by solar panels and stored in batteries, a task managed by URSC's power system engineers.

## 2. Vikram Sarabhai Space Centre (VSSC), Trivandrum:

- VSSC contributed to the thermal protection system for CH-3.
- They provided the pyro systems for deployment.
- VSSC played a crucial role in providing solar panel substrates for CH-3.
- Communication engineers at VSSC designed and tested advanced communication subsystems, ensuring seamless communication between CH-3 modules and ground stations.

## 3. Liquid Propulsion Systems Centre (LPSC), Trivandrum & Bangalore:

- LPSC contributed four propulsion systems to the CH-3 mission, including the L110 core liquid stage and the C25 cryogenic upper stage for the LVM3 launch vehicle.
- They designed the propulsion system fuels and navigation modules for the LVM-3 rocket.

## 4. ISRO Satellite Tracking Centre (ISTRAC), Bangalore:

- ISTRAC played a critical role in tracking and communication management, ensuring the safe operation of the spacecraft.

## 5. Space Applications Centre (SAC), Ahmedabad:

- SAC developed eight camera systems for CH-3.

- Their contributions include CH-3 Lander Imager cameras, Ka-band altimeter, hazard avoidance sensors, rover imagers and data processors.
- SAC also selected the landing site for CH-3.

## 6. Laboratory for Electro-Optics Systems (LEOS), Bengaluru

- LEOS contributed to the mission by developing cameras, sensors, NavCams and various other sensors.
- These sensors play a crucial role in navigation during the Lander's power descent phase.

## 7. ISRO Inertial Systems Unit (IISU), Trivandrum

- IISU designed and manufactured the Laser Inertial Referencing & Accelerometer Package (LIRAP) for the CH-3 mission.
- LIRAP is a navigation software that enabled autonomous descent and decision-making for the Lander.

## 8. Satish Dhawan Space Centre (SDSC-SHAR)

- SDSC-SHAR conducted integrated hot tests and launched the mission on July 14, 2023, at 2:35 pm.

## 9. National Remote Sensing Centre (NRSC), Hyderabad

- NRSC contributed through ground testing and provided references and aerial imaging for sensor testing.
- They analyzed lunar ejecta and estimated ejecta halo formation.

*Engineers Working*

### 10. Physical Research Laboratory (PRL), Ahmedabad

- PRL developed the Alpha Particle X-ray Spectrometer (APXS) payload for the rover.
- They also designed the Chandra's Surface Thermophysical Experiment (ChasTE) payload for the lander.
- These scientific payloads will help determine the chemical composition and temperature of the lunar surface.

## Collaborations and Partnerships

The success of Chandrayaan-3 was not solely attributed to ISRO but also to a network of private and public organizations and agencies that played an instrumental role in the mission:

1. **NewSpace India Limited:** Formed by the Government of India, NewSpace India Limited was responsible for producing, assembling and integrating the launch vehicle with the support of an industry consortium. It was a crucial part of the Chandrayaan-3 project.

2. **Indian National Space Promotion and Authorisation Centre (IN-SPACE):** Established by the government, IN-SPACE serves as a regulator for the space exploration industry and played a vital role in coordinating and overseeing various aspects of the mission.

3. **Indian Space Association (ISPA):** As an apex nonprofit industry body, ISPA focused on fostering successful collaborations within the private space industry in India and connecting it with government agencies including ISRO.

4. **SatCom Industry Association (SIA-India):** SIA-India, a non-profit organization, represented the interests of the space industry in India, promoting cooperation and collaboration with ISRO.

These organizations not only facilitated coordination but also contributed to the success of the Chandrayaan-3 project, showcasing the collaborative spirit that underpinned India's space exploration efforts.

There were some more Scientific and Engineering Teams and agencies acting as support systems for the mission:

1. **Indian National Space Promotion and Authorisation Centre (IN-SPACE):** As the regulator for the space

exploration industry, IN-SPACE played a critical role in overseeing and regulating various aspects of the Chandrayaan-3 mission.

2. **Hindustan Aeronautics Limited (HAL):** HAL supplied crucial components to National Aerospace Laboratories for launch vehicle testing, strengthening its long-term cooperation agreement with ISRO.
3. **Mishra Dhatu Nigam Limited (MIDHANI) Hyderabad:** With a partnership spanning over four decades, MIDHANI supplied essential products such as cobalt base alloys, nickel base alloys, titanium alloys and special steels, which were used in Chandrayaan-3 payloads.
4. **Ministry for Micro, Small and Medium Enterprises (MSME):** The Bhubaneswar Tool Room, a unit of MSME, manufactured around 54,000 aerospace components of 437 varieties for the Chandrayaan-3 mission.
5. **Institute for Design of Electrical Measuring Instruments (IDEMI) Mumbai:** IDEMI played a crucial role in manufacturing parts for Chandrayaan-3, contributing to its success.
6. **Semiconductor Laboratory (SCL) Mohali:** Under the Ministry of Electronics and Information Technology, SCL contributed to the fabrication of the Vikram Processor and Complementary Metal Oxide Semiconductor (CMOS) camera configurator for navigation and imaging.

7. **Kerala Minerals & Metals Ltd. (KMML):** A government-owned company, KMML supplied titanium sponge alloys for critical components in the Chandrayaan-3 mission.

8. **Kerala State Electronics Development Corporation Limited (KELTRON):** This public enterprise from Kerala played a crucial role in producing electronic power modules and the test and evaluation system for Chandrayaan-3.

The contributions of these agencies and organizations highlight the multidisciplinary and collaborative nature of Chandrayaan-3's success. Countless individuals, both in the limelight and behind the scenes, played crucial roles in the advancement of science and technology, demonstrating India's capabilities in the field of space exploration.

# Chapter 10

# India's Ascent to the Lunar Hall of Fame

## India's Quest to Join the Lunar Landing Elite

Once upon a time, in the heart of India, there was a quest, a dream that soared as high as the moon itself. It was the dream to land on the moon, to take a giant leap into the cosmic unknown and to etch India's name among the lunar landing elite.

India had always been captivated by the celestial dance of the stars. It had sent satellites into the vastness of space, exploring the mysteries beyond our planet's borders. But the moon, that silvery orb in the night sky, remained an elusive destination.

Then, in 2008, something incredible happened. India's space agency, ISRO, launched Chandrayaan-1, its first mission to the moon. Chandrayaan, which means "moon vehicle" in Hindi and Sanskrit, was about to take its first steps into the lunar world. The spacecraft went into orbit around the moon and made a groundbreaking discovery – it found water on the moon!

But Chandrayaan-1 wasn't the end of the story. India wanted to go even further, to achieve what only a select few nations had done before – a soft landing on the moon's surface. It was a tough journey, with many ups and downs, but the dream kept pushing forward.

## The International Impact and Collaborations

The dream of reaching the moon wasn't just India's alone. Space exploration has always been a global affair, a chance for nations to come together, share knowledge and explore the cosmos.

Chandrayaan-2, India's second lunar mission, showcased this spirit of international collaboration. It was a joint mission between ISRO and Russia's space agency, Roscosmos. The Russian lander added an extra layer of expertise to India's lunar dreams.

The journey didn't stop there. The moon's South Pole was a particularly challenging place to land on due to its rugged terrain and extreme temperatures. India wasn't alone in this quest. Other nations, like the United States and China, were also eyeing the lunar South Pole for exploration.

In 2019, India made an attempt, but the mission didn't quite go as planned. Yet, it wasn't the end. India remained determined and Chandrayaan-3 was born. It aimed to land on the moon's South Pole and it was part of a worldwide trend of lunar exploration. NASA had its Artemis III mission lined up for 2025, another significant step in the moon's southward journey.

## Inspiring Future Generations and Deep Space Exploration

The story of Chandrayaan-3 isn't just about landing on the moon. It's about inspiring future generations, the young minds who look up at the night sky and wonder about the mysteries of the cosmos.

The success of Chandrayaan-3 would light the way for aspiring scientists and space enthusiasts. It's a reminder that big dreams, no matter how challenging, can be achieved with dedication and teamwork.

India's lunar journey is just a stepping stone. It's a chapter in the grand story of space exploration. The moon is a testbed for what's to come – missions to Mars, asteroid mining, and deeper journeys into the galaxy. Chandrayaan-3 is part of that bigger story, a reminder that humanity's quest for knowledge knows no bounds.

So, as we look to the stars, let's remember that the dream to explore the cosmos, to unravel its secrets and to land on distant worlds, is a dream that belongs to us all. India's ascent to the lunar hall of fame is a testament to the spirit of exploration that unites people across the globe, from every age and background. The story of Chandrayaan-3 is still being written, with many more exciting chapters to come.

## Conclusion: A New Dawn in Lunar Exploration

And so, we reach the end of our lunar journey through the pages of Chandrayaan-3's remarkable mission. This story is more than just an account of a lunar landing; it's a testament to human curiosity, the spirit of exploration and the indomitable will to transcend the boundaries of our world and venture into the cosmic unknown.

When we look back on Chandrayaan-3 and its achievements, we see a mission that was about more than landing a rover on the moon's surface. It was about resilience and determination, about learning from setbacks and turning them into stepping stones. It's a testament to India's commitment to the advancement of science and technology and to its growing presence on the global stage of space exploration.

Chandrayaan-3 holds a special place in the annals of lunar exploration. It demonstrated that success is not always defined by a flawless landing, but by the pursuit of knowledge, the spirit of adventure, and the unyielding pursuit of scientific goals. The mission successfully carried out its scientific objectives, paving the way for future lunar endeavors.

But Chandrayaan-3 is just one chapter in India's ongoing journey to the moon. As you've seen throughout this book, India has embraced the challenge of lunar exploration with open arms and unwavering determination. The Indian

Space Research Organisation (ISRO) has proven itself as a formidable player in the global space community, leaving a distinct mark on lunar science.

Chandrayaan-1, India's first lunar mission, laid the foundation for future endeavors by discovering water on the moon. Chandrayaan-2, though it faced challenges during its landing, demonstrated India's capabilities in launching complex lunar missions. With Chandrayaan-3, the country has taken another step forward on its lunar quest.

ISRO's dedication to advancing space technology and exploration is palpable. With each mission, it has expanded its horizons and its goals. The moon is not the final destination; it is a stepping stone to the stars. The journey to the moon is not merely a quest for the lunar surface but a path to unlocking the secrets of our cosmos.

Chandrayaan-3 arrives at a time when lunar exploration is experiencing a renaissance. Nations and organizations around the world are turning their gaze toward our celestial neighbour, eager to uncover its hidden stories and unlock the doors to future exploration. The moon has become a hub of activity, not just for scientific research but also as a platform for international collaboration.

The moon represents an open playground for humanity. Its surface is a canvas on which we can paint our hopes and dreams for a better future. In this era of space exploration, the moon is at the forefront of our ambitions. Its resources, from water to minerals, promise new opportunities for scientific discovery and economic growth.

Chandrayaan-3 is part of this global wave of lunar exploration, but it is also a reflection of India's unique contribution to this endeavor. It tells the story of a nation with a rich history, diverse culture and a commitment to harnessing the power of science and technology for the betterment of humanity.

As we conclude this journey of the story of Chandrayaan-3, let's remember that space exploration is a shared dream that transcends borders and politics. It's about expanding the boundaries of human knowledge, discovering our place in the cosmos and shaping a brighter future for generations to come.

India's ascent to the lunar hall of fame is not the end but a new beginning. It is an invitation to dream bigger, reach higher and explore further. The lessons learned from Chandrayaan-3 will continue to inspire us, not only in our quest to understand the moon but also in our pursuit of the stars.

The mission's significance lies not only in its scientific achievements but also in the hope it instils in the hearts of people everywhere. It's a reminder that we, as humanity, are capable of great feats when we work together, learn from our past and aim for the skies.

The story of Chandrayaan-3 ends, but the story of lunar exploration continues. The moon, our celestial companion, remains a beacon of curiosity, a testament to human ingenuity and a symbol of our shared aspirations. It is a constant reminder that we are part of a grander cosmic narrative and our journey has only just begun.

As we close this chapter, we look up at the night sky and the silver orb that has sparked our imaginations for millennia. The moon, with its mysteries, its secrets and its allure, is not just a destination but a gateway to a universe of possibilities. So, let's keep looking up, dreaming and exploring. After all, the cosmos is ours to discover.

# Appendix

## (A Glossary of Lunar and Space Terminology)

1. **Cryogenics** - The study of the production and behavior of materials at extremely low temperatures to lift and place the heavier objects in space.

2. **Ejecta** - It is the matter that gets thrown out of a hole in the ground when a volcano erupts or when a meteorite crashes into the Earth.

3. **Ejecta Halo** - It is a pattern of rock, soil, and dust from the lunar surface that's thrown out in all directions around a crater, created by the impact of the ejecta.

4. **Electrojets** - They are electric currents that travel through the Earth's ionosphere. They are narrow belts of fast-moving ions.

5. **Elliptic Parking Orbit** - It is an oval-shaped orbit that is not a perfect circle. This means that the spacecraft is not always at the same distance from Earth. When it is closest to Earth, it is in its perigee and when it is farthest from Earth, it is in its apogee.

6. **Geosynchronous Transfer Orbit** – It is like a stepping stone in space where a satellite is put before it gets to its final spot. In this area, the satellite goes around the Earth at the same speed as the planet. It's like a midway point before it settles into a position above the Equator where it seems to stay still.

7. **Helium-3** - It is a type of helium that's lighter than the regular helium you might find in balloons. It has two protons and only one neutron, unlike other helium with two protons and two neutrons. It's the only stable form of any element that has more protons than neutrons, except for regular hydrogen.

8. **Ionosphere** – It is an upper atmospheric layer, 80-600 km up, where sunlight makes charged particles. It affects radio waves for communication and navigation.

9. **KeV** - It is a unit of energy that stands for kiloelectron volt. It's equal to 1,000 electron volts (eV).

10. **Laser-Induced Breakdown Spectroscopy** – It is a strong tool for finding out what stuff is made of. It can figure out what elements are in solids, liquids, or gases and it's good at getting rid of interference and finding weak signals without changing their strength.

11. **Liquid Apogee Motor** – It is a type of chemical rocket engine commonly used as the primary propulsion system in a spacecraft.

12. **Moon Impact Probe** – It was a 32-kilogram lunar probe that was released by India's Chandrayaan-1 mission. The

probe was designed to test systems for future landings and study the lunar atmosphere. It was the first to detect water on the Moon's surface.

13. **Natural gamma-rays** – They are electromagnetic radiations which are emitted when an unstable element (radioisotope) decays. They are imperceptible to human eyes.

14. **Payload** - It is an object or entity that is carried by an aircraft or launch vehicle. It can also refer to the carrying capacity of an aircraft or launch vehicle, which is usually measured in terms of weight

15. **Propulsion** – It is the act of moving or pushing an object forward. The word comes from two Latin words: 'pro', meaning before or forward, and 'pellere', meaning to drive.

16. **Regolith** – It is a layer of loose material that covers solid rock. It includes dust, broken rocks, soil and other related materials. It is present on Earth, the Moon, Mars, some asteroids, and other terrestrial planets .

17. **Solar Neutrons** – They are tiny particles made during solar flares. They interact with hydrogen in the Sun's surface and release high-energy rays in space.

18. **Stratigraphy** – It is a branch of geology that studies rock layers, or strata. It's also known as historical geology.

19. **Sun Synchronous Orbit** – It is an orbit that goes around a planet near its poles. In this orbit, a satellite flies over

the same spot on the planet's surface at the same time every day.

20. **Translunar** – It means something is towards the moon from Earth or another planet.

21. **Velocimeters** – A velocimeter is a tool that measures how fast something, like an object or a wave, is moving. Velocity means how quickly something is going from one place to another.

22. **X-Ray Astronomy** – It is a type of astronomy that looks at X-rays coming from things in space, like stars, hot gases and galaxies. It helps astronomers learn about the universe, like why young stars give off more X-rays than older ones.

# References


https://www.isro.gov.in/Chandrayaan3_Details.html

https://www.isro.gov.in/Making_Chandrayaan3_ISRO_culture.html#:~:text=Major%20scientific%20experiments%20planned%20once,conductivity%20up%20to%20the%20depth

https://www.bbc.com/news/world-asia-india-66808808

https://blog.jatan.space/p/why-explore-the-moon

https://www.isro.gov.in/Chandrayaan_2.html

https://www.isro.gov.in/Launcher.html